# 50 Perspectives of the Cosmos

# 50 Perspectives of the Cosmos

One 14 Year Old's Thoughts on Our World

LUKAS WINKELMANN

Copyright © 2024 Lukas Winkelmann

All rights reserved. No part of this book may be reproduced in any form or by any means, electronic or mechanical, including information storage and retrieval systems, without permission from the author or publisher.

ISBN: 9798327085084

Cover and book design: Lukas Winkelmann
Tree of Discovery cover photo: Lukas Winkelmann
Milky Way cover photo: Adrian Pelletier - unsplash.com

For all those who are curious

with the desire to explore...

# Table of Contents

An Introduction to Our World Page 1

Part One:
PERSPECTIVES OF SCIENCE Page 5

PERSPECTIVE ONE: Page 7
Why is it Important to Imagine?

PERSPECTIVE TWO: Page 11
Why Should We Be Curious?

PERSPECTIVE THREE: Page 13
What is Truth?

PERSPECTIVE FOUR: Page 14
What is the Scientific Method?

PERSPECTIVE FIVE: Page 17
Why Does Science Exist?

PERSPECTIVE SIX: Page 19
Is Science Hidden From Us?

PERSPECTIVE SEVEN: Page 20
How Did Exploration Begin?

PERSPECTIVE EIGHT: Page 22
Should We Ask Questions?

PERSPECTIVE NINE: Page 23
How Stable is the Scientific Community?

PERSPECTIVE TEN: Page 25
Why Do Science and Religion Frequently Seem to be Considered a Conflict of Interest?

# TABLE OF CONTENTS - CONTINUED

Part Two:
**PERSPECTIVES OF THE UNIVERSE** Page 29

**PERSPECTIVE ELEVEN:** Page 31
What is the Definition of the Universe?

**PERSPECTIVE TWELVE:** Page 32
Does Everything in the Universe Have to Have Matter?

**PERSPECTIVE THIRTEEN:** Page 34
How Do We Measure the Vast Reaches of the Universe?

**PERSPECTIVE FOURTEEN:** Page 35
What is an Astronomical Unit (AU)?

**PERSPECTIVE FIFTEEN:** Page 36
How Do We Use Scientific Notations?

**PERSPECTIVE SIXTEEN:** Page 37
What is the Medium of Sound?

**PERSPECTIVE SEVENTEEN:** Page 38
Does Light Have a Medium?

**PERSPECTIVE EIGHTEEN:** Page 39
Does Air Have Weight?

**PERSPECTIVE NINETEEN:** Page 40
Are Vacuums Possible?

**PERSPECTIVE TWENTY:** Page 42
Do We Live in a 4D Universe?

**PERSPECTIVE TWENTY-ONE:** Page 43
Would We Exist Without Time?

## Table of Contents - Continued

**Perspective Twenty-Two:** Page 44
What is the Fastest Speed in the Universe?

**Perspective Twenty-Three:** Page 45
What Was the Discovery of Special Relativity?

**Perspective Twenty-Four:** Page 47
What is the Stubborn Illusion or Dilation of Time?

**Perspective Twenty-Five:** Page 49
What Happens to Time at the Speed of Light?

**Perspective Twenty-Six:** Page 50
Do We Ever Stop Moving?

**Perspective Twenty-Seven:** Page 51
How Does Time Dilation Work in Our Daily Lives?

**Perspective Twenty-Eight:** Page 53
What is Gravity?

**Perspective Twenty-Nine:** Page 55
What is General Relativity?

**Perspective Thirty:** Page 57
What is Spacetime?

**Perspective Thirty-One:** Page 59
What is the Gravitational Time Dilation?

**Perspective Thirty-Two:** Page 61
Is Time Travel Possible?

## Table of Contents - Continued

**Perspective Thirty-Three:** Page 62
What is a Black Hole?

**Perspective Thirty-Four:** Page 64
What is the Quantum World?

**Perspective Thirty-Five:** Page 66
What is Quantum Mechanics?

**Perspective Thirty-Six:** Page 68
What is Inside a Black Hole?

**Perspective Thirty-Seven:** Page 70
Is There a Difference Between Quantum Mechanics and General Relativity?

**Perspective Thirty-Eight:** Page 72
What is String Theory?

Part Three:
**Perspectives of Our Planet** Page 75

**Perspective Thirty-Nine:** Page 77
Why Do So Many People Respect the Stars?

**Perspective Forty:** Page 78
Do Our Eyes See Everything?

**Perspective Forty-One:** Page 79
Why Do Events Happen?

**Perspective Forty-Two:** Page 80
How Advanced is Our Civilization?

## Table of Contents - Continued

**PERSPECTIVE FORTY-THREE:** Page 81
What is the Survival Rate of the Human Species?

**PERSPECTIVE FORTY-FOUR:** Page 82
How Much is the Earth Cared For?

**PERSPECTIVE FORTY-FIVE:** Page 83
What is Earth Day?

**PERSPECTIVE FORTY-SIX:** Page 85
How Much Impact Does the Earth Have on Our Lives?

**PERSPECTIVE FORTY-SEVEN:** Page 87
What is Living, and What is Not?

**PERSPECTIVE FORTY-EIGHT:** Page 88
Our World in the Present.

**PERSPECTIVE FORTY-NINE:** Page 90
Lessons Taught and Lessons to be Prepared.

**PERSPECTIVE FIFTY:** Page 97
The Scientific Tree of Discovery

**EPILOGUE** Page 102

**ABOUT THE AUTHOR** Page 107

# Acknowledgments

First, I would like to express my tremendous gratitude to my teachers, who have supported me throughout this journey of writing and publishing a book. Special thanks to my inspirational science teachers, Roger McElmell for Physics, Ty James for Chemistry, Al Smith for Biology, my Cohort and Math teacher, David Morningstar and Jonathan Sonoda for publishing advice. I appreciate my 1st Grade teacher, Laura Slough, who has remained in contact with me all the way into High School and has also spread the word of this book to others she knew. Also, a huge thanks to my World History teacher, Nicholas Heyming, for reaching out way beyond and putting me in contact with others.

My appreciation to Ben Haldeman, who read my book and offered me the incredible opportunity to send this book to the moon aboard a Firefly Blue Ghost lunar lander planned to launch on a SpaceX Falcon 9 rocket in October 2024.

Finally, my utmost gratefulness to my Mom and Dad. Thank you for being such an incalculably boundless support throughout this marvelous journey of mine.

# An Introduction to Our World

We are the people of planet Earth... intelligent, brave, and adventurous. This planet is what we may always call home.

Think of all our politics, religions, sciences, and more. Our species appears to be beaming with life and many more significant possibilities.

Our personalities depend on our moods. Our moods depend on how we relate to one another. Relationships are critical to our species' survival.

The human industrialization of our world changed the way we existed more than a century ago. Many extraordinary lives have been saved since then. Lives that may have never had a chance to live until now.

No matter your race, your religion, or the country you were born in, we are all people, just with different manners. Each culture displays kindness differently. So, why do we still argue about this and can't just get along?

Perhaps the answer lies right in front of us. Humans of different societies have only sometimes been presented to other cultures. Therefore, we get offended when we see different mannerisms because we think they are a sign of disrespect. However, to other cultures, it is a sign of respect. Relationships with societies worldwide have been disparaged because of the misinterpretation of one another.

Of course, we are all biased to some of our own beliefs and

traditions. We all have different ways of showing our thoughts and manners, but they result in the same demeanor to express kindness...

Before we begin to study science, let's put aside labeling people's political and religious beliefs. It's all right to commit to your beliefs and not change them. But, if we continue to label each other because of them, we are certainly not on the road to success.

Luckily, we are not near that far yet. We, as people, are constantly changing. All we need is a good chat with each other. Then, we can finally obtain a global agreement, to maintain individuality, and a great population variety without violence.

This book is a brief explanation of perspectives on our world and science. Science was not designed to be a political or religious thing. It was instead created by curious minds who acquired an interest in exploring.

As a result, science can be a safe way to communicate, which global nations must agree upon.

Science is constantly changing, so it is essential not to be offended if your discovery, for instance, is disproved because of such a fluctuating matter.

So, what is the definition of science? What is its meaning? Why does it exist?

Science is defined as the systematic study of the behavior and structure of the natural universe. It is proved by observation and experimentation.

But really, science is like the combination, or family tree, of people with curious minds, willing to share their observations, later to be proven, then corrected by future generations who maintain the same curiosity.

In this book, we will witness perspectives on science that may help instigate your new ideas that may one day form discoveries. Perhaps anyone can be a scientist at any age, if they are curious and eager to explore.

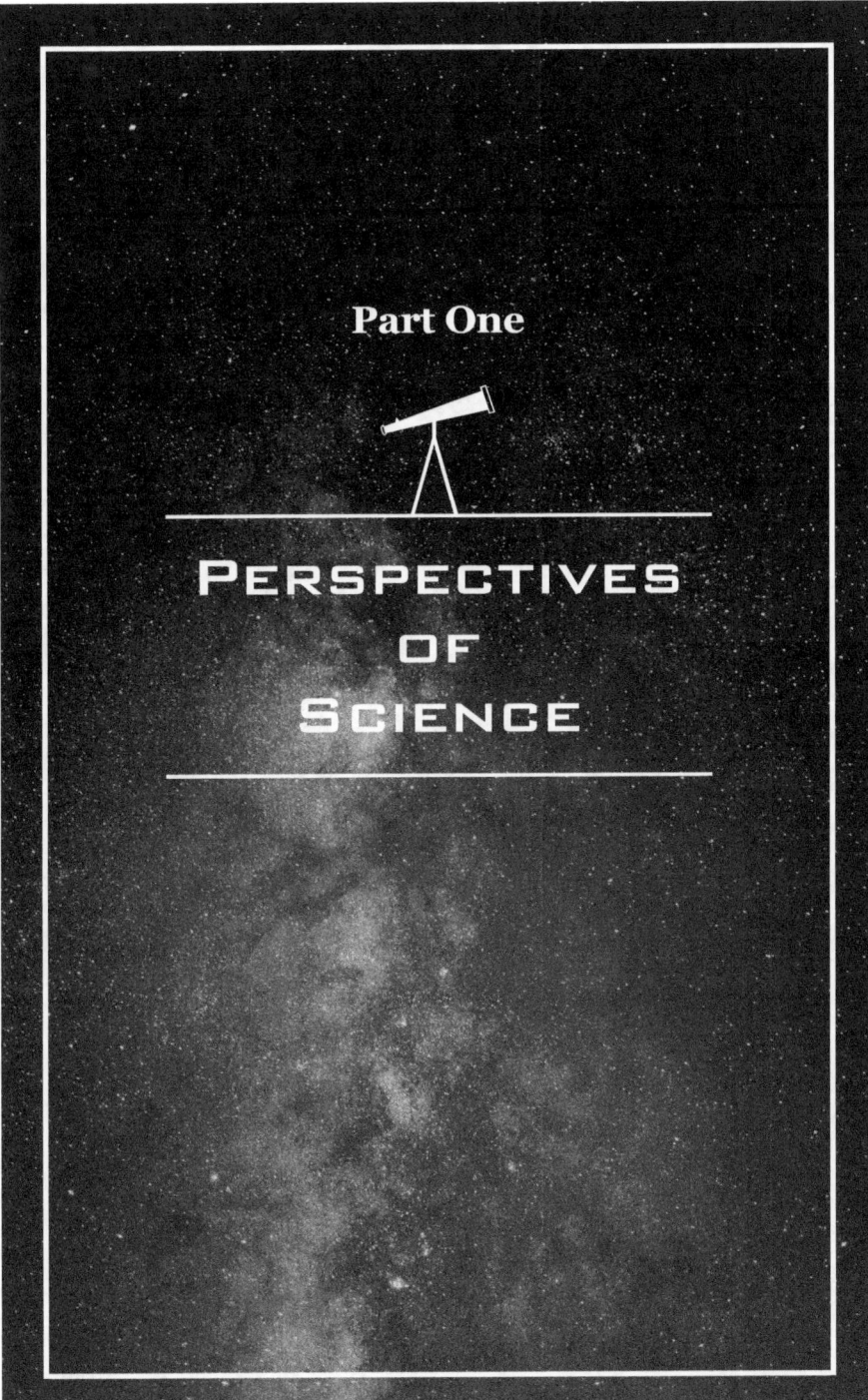

## Perspective One

# Why is it important to imagine?

*"Imagination is more important than knowledge."*
-Albert Einstein

Before embarking on our journey through the vast expanse of the universe, it is essential to take note of the key elements that have paved the way for numerous scientific breakthroughs. Why do we begin here?... Well, science has more or less been a stumbling journey for countless years, full of thoughtworthy ideas that were much denied during their first introductions.

When Albert Einstein, a famous scientist of the late 19th century, and early 20th century, published what is called his Special Theory of Relativity in 1905, it was met with mixed reactions. Although some scientists were curious about his ideas, others were skeptical. The scientific community took some time to fully understand and appreciate the revolutionary nature of Einstein's ideas. Initially, some physicists who were more familiar with classical Newtonian physics, which was the science developed by Sir Isaac Newton in the 17th century, 300 years before, were critical of Einstein's theory. The concepts introduced by Einstein were radically different from what was previously understood about space and time.

However, as time passed, more and more experimental evidence began to support the predictions made by special relativity. One such example is the famous Michelson-Morley experiment, which is discussed later in this book. The experiment, conducted in the late 19th century, failed to detect any "aether," which was previously believed to be the medium through which light traveled. Special relativity provided an explanation for this result.

Moreover, technological advancements enabled scientists to conduct experiments and make observations that confirmed different aspects of special relativity. For example, particle accelerators and high-precision atomic clocks provided evidence that supported the theory's prediction of time dilation.

It's important to note that not every new idea is immediately accepted when it's published. Breakthroughs often take time, research, and personal experimentation before enough people begin to realize their true meaning.

However, when an idea is finally proven by more research, the world may have an entirely new perspective of the universe...

Keeping in mind the important facts about how an idea can evolve into a discovery, let's take a step back and consider the key elements that have led to numerous scientific breakthroughs in the past. Undeniably, the first element is imagination.

Imagination is a crucial component of both creativity and science. It is closely tied to creativity, but in order to be creative, one must first have an active imagination.

Then, what does creativity even have to do with science?

Creativity opens up the mind to explore fictional universes, which can lead to unconventional and thought-provoking ideas. These ideas, though they may seem strange compared to current understanding, can inspire new and innovative experiments. Even if the experiment doesn't produce the desired result, it's still a discovery in its own right. True discovery in experimentation is not only about achieving success, but also discovering what doesn't work. If an experiment doesn't work out, it's an opportunity to try a different approach.

Each time an experiment doesn't succeed, it brings you one step closer to finding something that does. After conducting many different experiments, there will come a time when the results seem to have a positive outcome. When this happens, it's important to repeat the exact same experiment again and again, to confirm the consistency of the results.

Once you've repeated the experiment enough times and have confirmed the consistency of the results, it's time to share your new idea with other scientists. Similar to Einstein, you may encounter individuals who challenge your claim and point out flaws in your experimental procedure, potentially leading to the rejection of your idea. However, there are instances where an idea can be replicated across the world with similar results, providing certainty that the idea is correct. When that day comes, it marks an advancement in the global scientific community. If, after a few centuries, a new idea emerges that contradicts your discovery, as has happened to some of Isaac Newton's, it means that others have built off of your discovery, leading to new breakthroughs that shape our understanding of the universe.

*Even some of the most influential innovations of our time, would likely not have become a reality without one's great imagination.*

Photo source: Unsplash.com

## Perspective Two

## Why should we be curious?

*"The important thing is not to stop questioning. Curiosity has its own reason for existing."*
-Albert Einstein

Every child is born with a natural sense of curiosity, and some individuals carry this trait into adulthood. Such people are like travelers, always seeking to learn and explore the world around them. Self-teaching is one of the most important types of learning, where you become your own teacher. You generate questions that you are eager to answer, some of which may already have answers, while others may not. By asking these questions, making hypotheses, and investigating the topic, you can become the first to discover something new.

Sometimes, this curiosity vanishes when people grow up. However, perhaps it can always be rebuilt by anyone at any age.

The steps to rebuilding curiosity are straightforward. Remember to stay inquisitive and never stop exploring.

Think of a scientist as a grown-up child who never lost their sense of curiosity. If you too have managed to preserve this curiosity, then you have already embarked on a journey of exploration, mentally venturing out into the vast expanse of the universe.

## Perspective Three

## What is Truth?

*"The opposite of a fact is falsehood, but the opposite of one profound truth may very well be another profound truth."*
-Niels Bohr

Is truth a hunt, or is it a belief? There are many who enjoy debating the truth, but determining what is true can be a complex process. This is because each person may have their own interpretation of the facts, making the concept of truth an ongoing and never-ending debate in the pursuit of greater knowledge.

Knowledge is always evolving, which means that we can never be entirely certain that we are correct. The truth is a constantly changing concept.

However, we must all agree on one thing if we are to seek the absolute truth: if a fact remains a topic of debate, then we simply do not know, or have enough evidence about it yet...

## Perspective Four

# What is the Scientific Method?

*"I am not accustomed to saying anything with certainty after only one or two observations."*
-Andreas Vesalius

If you have ever conducted an experiment or made a discovery, there's a good chance you used what is called the scientific method without even realizing it. For most people, including scientists, it's difficult to avoid using this method because it provides a clear sequence of steps that make the experimental process much more manageable.

We'll take a closer look at those steps to gain a better understanding of the experimental process. We'll explore the eight primary stages of the Scientific Method.

Let's start with the first step, which is called "Observe." It's one of the most crucial stages of the scientific method because you can only formulate a question if you observe something first. To begin the scientific method, you need to be curious and never stop exploring your surroundings. Who knows, you might observe something that instigated your interest one day...

The following step is called "Question."

After making an observation, it's crucial to ask yourself a question. For instance, suppose you're a professor at a university, and one of your students asks you a question about planetary exploration. As a planetary scientist, it's your responsibility to answer that question. The same applies to the scientific method. You'll be the one asking the question, but you won't have an answer yet.

With your curiosity and desire to learn, you'll define the question and start researching the topic. This step is, of course, called "Research."

Conducting research enables you to uncover existing answers and information on the topic. This step can help determine if someone else has already found a solution to your question or if there are any existing records of past recognition on the subject.

If you're fortunate, you may find an easy answer to your question on the modern internet. However, if you're not, you'll need to conduct more research. If you don't find any solutions but gather information on the topic, it's time to move on to the next step of the scientific method, which is "Formulating a Hypothesis."

A hypothesis is a proposed explanation based on limited evidence from previous research. It serves as your starting point for further investigation. As we move through the next steps of the scientific method, your hypothesis may be proven or disproved.

Once you have your hypothesis, it's time to move on to the step which is "Experiment." At this stage, the scientific method allows you to develop a procedure to test your hypothesis. The experiment can be anything that tries to prove your hypothesis. However, it must be accurate and obtain actual data.

After the experiment, it is time to collect the data at which you obtained. Once you have your data, it's time to move onto another step, which is "Analyze and Explain Results." This is where you'll likely make new discoveries, whether or not your hypothesis is proven.

Finally, you can conclude your experiment and compare the data you've collected to your hypothesis.

After repeating the experiment many more times with the same results, you can reveal and report your results to the scientific community. This is where scientists worldwide will also conduct the same experiments to prove whether your discovery is indeed a reality. This concludes the scientific method.

The scientific method is used by renowned scientists worldwide, and it has helped reveal many aspects of our surroundings. If we can all learn to use the scientific method, we can make rapid advancements. Imagine how much more advanced our daily lives could be if everyone observed and questioned our surroundings. Every person could make groundbreaking discoveries that could improve everyone else's lives like never before.

## PERSPECTIVE FIVE

## WHY DOES SCIENCE EXIST?

*"Equipped with his five senses, man explores the universe and calls the adventure Science."*

-EDWIN HUBBLE

Throughout the book, we have explored different viewpoints that are rooted in science and its impact on society. However, one important aspect that we haven't touched upon yet is the reason behind the very existence of science.

One possible reason for the existence of science is that it serves as a tool for society to gain a better understanding of the world around us. By observing and analyzing the present, we can make predictions about the future. Additionally, science allows us to delve into the past and learn from it, using common scientific laws and remaining clues from ancient times.

Although it is difficult to imagine what our society would be like without science, it is clear that science has had a significant positive impact on our lives. Through advancements in medical research, science has helped us lead healthier lives. Furthermore, it has made our surroundings more enjoyable by pushing the boundaries of what we thought was impossible.

*The existence of science has allowed us to travel beyond what we knew. In this photo, we experience the Earth from above traveling aboard a spacecraft. Below, we see all the fascinating connectivity of roads and cities. These cities and lights to which you see would not have existed if it were not for the inventions founded off of scientific ideas. Science has allowed us to be more connected.*

Photo source: Unsplash.com

## Is Science Hidden From Us?

*"What goes on inside a star is better understood than one might guess from the difficulty of looking at a little dot of light through a telescope because we can calculate what the atoms in the stars should do in most circumstances."*

-Richard Feynman

In the world of science, there's a unique challenge that we face, especially when it comes to the microscopic realm of atoms. It's not easy to observe this hidden world with our naked eyes. Imagine how much more advanced our society could be if we could witness science in action at all levels! But how do we know about the existence of these things that are beyond our vision?

The answer lies in mathematics. Many people consider math to be the language of the universe as it holds the key to unlocking the mysteries of the cosmos. By mastering this universal language, we have the potential to calculate and predict the workings of the entire universe.

## Perspective Seven

## How Did Exploration Begin?

*"I think of space not as the final frontier but as the next frontier. Not as something to be conquered but to be explored."*

−Neil DeGrasse Tyson

In the beginning, the first humans walked the Earth, and yet so much of their journey remains a mystery. However, one undeniable fact remains - they achieved something truly remarkable. They stumbled upon the vast shore of the Earth's great ocean and set out to explore what lay beyond. The story of their bravery and determination still astonishes the world today.

Their first accomplishment was the construction of something that could float on water - what we now call boats. With these vessels, they set sail into uncharted territories, bravely venturing out into what they believed was a flat world. They sacrificed the comfort of their homes to embark on a journey towards the unknown, with no clear understanding of what lay ahead of them.

Their efforts eventually paid off as they found new lands and mapped the world, which changed the course of human history forever.

Today, we find ourselves in a similar scenario yet again. We are exploring a new shore that leads to the stars, venturing into the unknown with the same sense of curiosity and bravery. It's amazing to think about how much we have accomplished and how much more we have yet to discover. The journey towards the stars is fraught with challenges, but with our collective determination and courage, there's no telling what we can achieve in the years to come.

*Ships, similar to the one in this photo, signify humanities ability to explore new, unknown territories.*

Photo source: Unsplash.com

# Perspective Eight

## Should We Ask Questions?

*"Learn from yesterday, live for today, hope for tomorrow. The important thing is not to stop questioning."*
-Albert Einstein

Have you ever wondered whether asking questions is important? The answer is a resounding "Yes!" In fact, asking questions is an effective way to demonstrate your interest and enthusiasm in a particular subject. It shows that you are willing to learn and understand more about the topic at hand. Moreover, asking questions helps us to delve deeper into the subject matter, enabling us to correct any errors we may have made and gain a greater understanding of the topic.

Furthermore, asking questions can also lead to unexpected discoveries and insights that we may have otherwise missed. As we question and seek answers, we are more likely to uncover new information and perspectives, leading to a more comprehensive understanding of the topic.

Therefore, it's important to never shy away from asking questions. Asking questions can only help us become smarter, more insightful, and more knowledgeable individuals.

## Perspective Nine

# How Stable is the Scientific Community?

*"Look up at the stars and not down at your feet. Try to make sense of what you see and wonder."*
-Stephen Hawking

Age is never a barrier to discovering something new, be it at the tender age of 2 or 102. The path to discovery is paved with curiosity, as we've seen before.

We are born with an innate sense of curiosity, and from the moment we can walk, we become natural explorers. However, our explorations are limited to our immediate surroundings, such as our home, backyard, and local city. But what about the third dimension that leads "up"?

As children start exploring the world beyond the ground level, they become fascinated by the mystery that lies above the earth. Sadly, as they grow up, this sense of wonder sometimes fades away, and they become more focused on the mundane aspects of their lives.

One of the reasons for this is the modern world we live in, where we are surrounded by towering skyscrapers and artificial lighting that obscures our view of the natural sky. Light pollution has become a significant problem in highly industrialized areas, and it's getting worse every day. The more artificial light we use, the fewer stars we can see.

As the global population continues to increase, so does the demand for more housing, which leads to more artificial lighting and more light pollution. There could be a point where children will never get to see the stars and experience the wonder of the universe.

This would be a devastating loss, not just for children but for everyone. Scientists need the sky to study the universe, and without that, human progress would come to a screeching halt.

But there is always hope with anything. We need to take a global approach to this problem and come up with innovative solutions that can help us conquer this challenge. There are already some ideas out there, but that does not mean that we do not have to continue researching and exploring new avenues to find a long-term, cheaper, solution.

## Perspective Ten

# Why Do Science and Religion Frequently Seem to be Considered a Conflict of Interest?

Science should not be involved with political or religious dogmas. However, in contemporary times, it seems to have become entangled in both domains. It is crucial to acknowledge that beliefs are an integral part of our identity, connecting us to our family and cultural heritage. Beliefs, more or less display who you are.

Science, on the other hand, does not have the power to alter one's representation in the world. That said, it is crucial to remain open-minded and curious. Even if you do not subscribe to the scientific method or desire to become a scientist, conducting experiments and exploring your surroundings can yield insights that might help improve our world. So, keep an open mind, be willing to learn, and together, we can create a better future.

## 🔭 Reflection

Before you continue, stop reading momentarily and become curious to explore your surroundings. Take note of your experiences, even on the most minor level. Do some of your experiments. Declare some of your hypotheses. See what information you gather in your area, then formulate a mental analysis. Make sure to be accurate in your comment.

Train your mindset to believe that you are searching for the validity of your analysis.

Once you're finished, share your analysis with people you know. If you were accurate, and the people you met were also accurate, then you will all receive equal results in the analysis of the experiment.

If your results differed, then one of you must need to follow the experiment more accurately. Here, it becomes essential that everyone retakes the investigation and more closely studies the data. Then, finally, if everyone received the same results and made an equal analysis, it becomes highly likely that the experiment was a success.

To be accurate, experiment for a final time and study the investigation even closer. If everyone still receives the same results, you can finally declare your experiments to the public, where people worldwide will accurately repeat the experiment and receive the same results.

If the global community receives equal results, the results can finally be declared a discovery. And the truth becomes more available.

The continuation of Perspectives in this next part of this book will be discoveries that will hopefully ignite your curiosity.

There may also be speculation in the upcoming perspectives. It is all right to speculate, as it is a delicate facet of imagination, creativity, and curiosity. Science would indeed not exist without these qualities.

However, we will be sure to mark the difference between the fictional context of our world and the truth. That way, we will understand.

If we memorize this difference between fiction and non-fiction, we have finally opened the doors to planetary exploration. Our universe can now be wondered and explored...

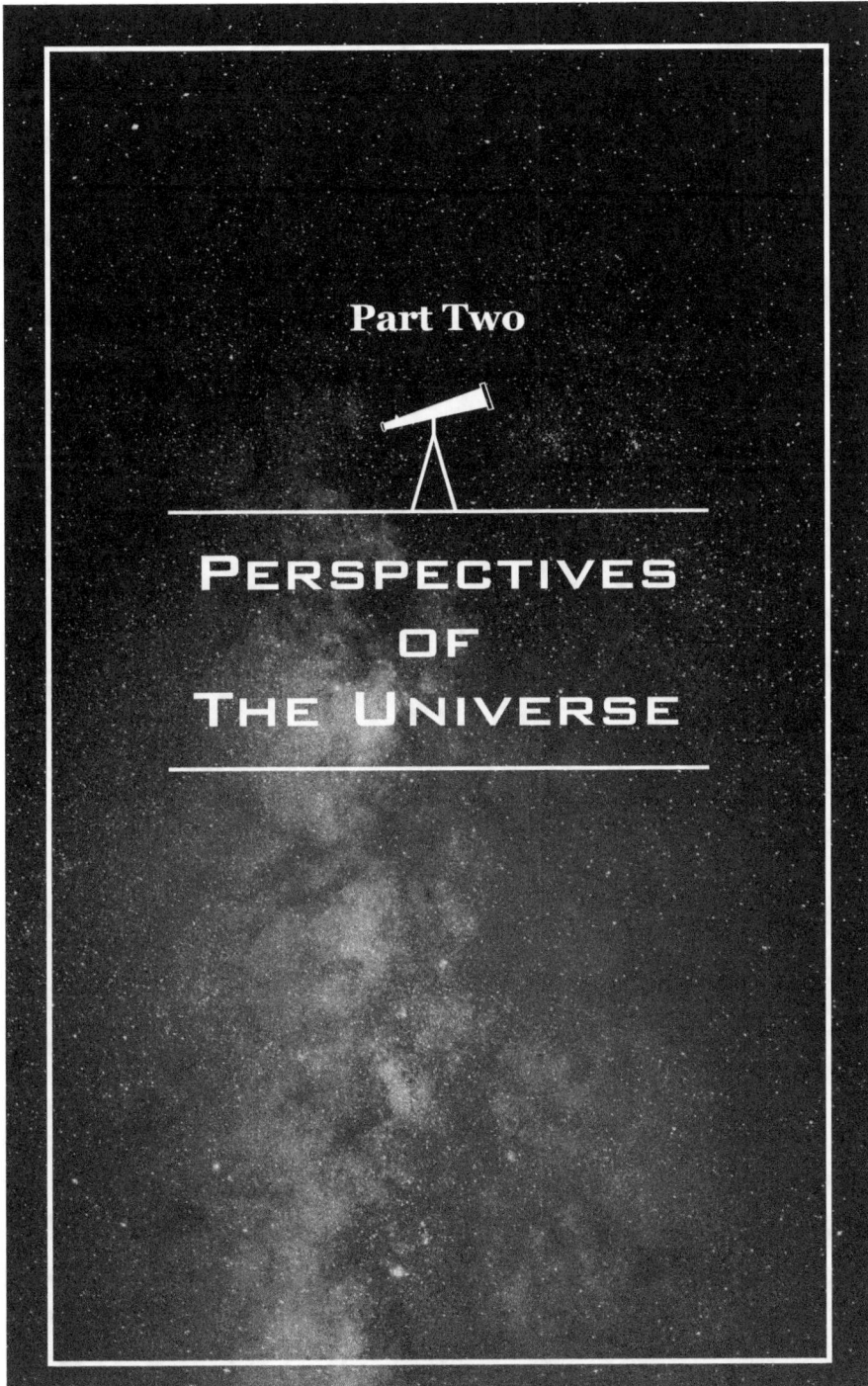

## What is the Definition of the Universe?

*"The total number of stars in the universe is greater than all the grains of sand on all the beaches of the planet Earth."*
-Carl Sagan

The Universe is vast and mysterious, and exploring its depths is a journey like no other. It's the story of everything, containing all that we know and so much more. But where do we even begin on this unimaginable platform?

Scientists have been pondering this question for centuries, and we've come a long way in our understanding of the Universe. We know that it's home to stars, planets, matter, elements, light, atoms, particles, time, and dimensions. But there's still so much we don't know.

How can we even begin to explore something that may be infinite?

Despite our limited understanding, we all have our own ideas about what the Universe means and what its purpose might be. And that's the beauty of it - the Universe remains undefined, waiting for us to uncover its mysteries and reveal its true purpose. So let's keep searching and exploring, and who knows what we may find among the stars.

## Perspective Twelve

# Does Everything in the Universe Have to Have Matter?

Matter is the fundamental building block of our physical world, encompassing any substance that takes up space by having volume. From the objects we use daily, to the vast expanse of stars and planets, matter is ubiquitous in our universe. Even the smallest particles, such as atoms, possess this essential quality. However, the question arises - is matter a prerequisite for existence? Is it possible that the universe contains entities that are void of matter?

The concept of non-material existence may seem unfathomable, as everything we perceive in our daily lives is tangible and has mass. But, what if mass is merely a term for something we can touch? What if the universe holds greater mysteries that challenge our understanding of matter? Perhaps these are some of the most profound questions we can ask about the cosmos. Could sound and light be considered matter? What about our thoughts? Perhaps not everything that exists requires matter to exist.

As we delve deeper into the mysteries of our universe, we are forced to question and challenge our fundamental assumptions about what defines reality.

*This photo, taken by the NASA Hubble Space Telescope, portrays a vast arena of galaxies, ones to which are millions of light years away. Each, containing huge bundles of matter and energy.*

Photo source: Unsplash.com

# How Do We Measure the Vast Reaches of the Universe?

The sheer vastness of the universe leaves us in awe and wonder. Measuring its immensity is a task that is hard to fathom. However, scientists have come up with a groundbreaking invention known as the light year. This unit of measurement is derived from the speed of light, which travels at an astonishing 299,792,458 meters per second. The light year provides a convenient and practical way of measuring the immense distances within the universe.

For instance, the distance between the sun and the Earth is 8.3 light minutes. This means that it takes 8.3 minutes for light to reach the Earth from the sun. When we consider the size of the Milky Way galaxy, we realize that it is on an entirely different scale. The Milky Way spans approximately 100,000 light-years across, which is most definitely beyond our comprehension. The light year helps us to grasp the immense size of the galaxy and the universe as a whole.

## Perspective Fourteen

# What is an Astronomical Unit (AU)?

The solar system, with its vast expanse, is a marvel of astronomical proportions. To put its size into perspective, scientists have devised a unit of measurement known as the astronomical unit (AU). This unit measures the distance between celestial objects in the solar system, with one AU equivalent to the distance between the Earth and the sun, spanning an impressive 150 million kilometers. To add to the nature of this distance, light takes approximately 8.3 minutes to traverse one AU, highlighting the immense scale of our solar system.

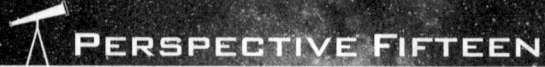

## PERSPECTIVE FIFTEEN

## HOW DO WE USE SCIENTIFIC NOTATIONS?

The magnitude of the universe is truly staggering, with dimensions so vast, and so small, that even the light year proves insufficient. Consequently, scientific notations are employed to fathom its size. For instance, the Milky Way Galaxy spans a distance of about 100,000 light-years, which can be more conveniently expressed as $1 \times 10^5$ light-years.

When dealing with even more colossal measurements, the Observable Universe spans a massive 92 billion light-years. This scale can be better comprehended as $9.2 \times 10^{10}$ light-years. Such scientific notations enable us to grasp the enormity of the cosmos and the mysteries that it holds.

## Perspective Sixteen

# What is the Medium of Sound?

Sound waves are a fundamental concept that is often taught in middle school science class. It is commonly understood that sound waves are measured by their frequency and amplitude, and that they propagate through a medium. However, in the realm of physics, sound is defined as a vibration that we hear, which propagates as an acoustic wave through a solid, liquid, or gas medium.

The process of hearing sound involves the vibration of molecules, which push off each other until they reach our eardrums. The eardrum then transmits these vibrations to our brain, which interprets them as sound. It is noteworthy that sound travels faster through liquids than in gases, and it travels the fastest through solids. However, sound propagation speed is still extremely slower than that of light.

## Perspective Seventeen

# Does Light Have a Medium?

The early physicists proposed that light waves require a medium similar to sound waves. This medium, known as the luminiferous aether, garnered immense attention and curiosity, yet remained an enigma to the scientific community. This concept, though plausible at the time, confused several scientists and left a significant gap in their understanding of the universe. The notion of an absolute aether implied that the vacuum of space was impossible, which further fueled the curiosity of scientists. However, with the emergence of new scientific knowledge and advancements, the aether's validity became a topic of debate. This debate culminated in the Michelson-Morley experiment of 1887, which determined that light always traveled at the same speed, regardless of its direction. This discovery suggested that the aether was a mere illusion.

## Perspective Eighteen

# Does Air Have Weight?

One may perhaps have pondered whether air possesses weight, and if so, why it does not seem to be visible to us. Interestingly, there exists a method for determining the weight of air that can be conducted through a simple experiment using a smooth, solid surface, such as a marble kitchen counter and a suction cup. The suction cup, when firmly pressed onto the counter top, becomes difficult to remove and must be peeled off. However, the suction cup is not functioning by suction at all. Rather, the difficulty in removing it is due to the creation of a vacuum-like effect inside, where all the weight of the air pressing down from above is concentrated. Given that air is present in the earth's atmosphere and beyond, it acquires a significant amount of weight. For instance, at sea level, one square inch of air is estimated to exert a weight of roughly 15 pounds in all directions. Therefore, if a suction cup with precisely one square inch of surface area were pressed onto a wall or counter top, it would be like lifting a weight of 15 pounds to dislodge it!

# Perspective Nineteen

## Are Vacuums Possible?

A vacuum is a state characterized by the complete absence of any form of matter. Even sound waves are unable to propagate through a vacuum due to the lack of a medium. In the past, creating a vacuum was deemed impossible by many scientists due to the weight of air, which seemed to crush everything in its path.

At sea level, air weighs 15 pounds per square inch, making it impossible for thin plastic straws to handle the weight when placed in a vacuum. However, with increasing curiosity and interest in the subject, new vacuum gadgets were designed to withstand this immense weight.

One obstacle that scientists faced in their quest to create a vacuum was the question of light. At the time, physicists believed that light waves required a medium, much like sound waves, known as the luminiferous aether. This presented a conundrum, as a vacuum could not be considered a vacuum if light required a medium to travel.

Thankfully, in 1887, physicist Albert A. Michelson and his colleague Edward W. Morley postulated that light did not require a medium to move through. This discovery finally lent credence to the accuracy of vacuum experiments.

However, further research revealed that space, which appears to be a vacuum, is not entirely devoid of energy or atom-sized particles. As a result, scientists continue to explore new ways of creating vacuums to this day.

## Perspective Twenty

# Do We Live in a 4-Dimensional Universe?

We are conventionally taught that the universe we inhabit is a three-dimensional space. Our mobility is restricted to moving up, down, left, right, forward, backward, and so on. But what about the fourth dimension? Can we conceive of a world that exists beyond the three dimensions we know?

The challenge we face in comprehending the fourth dimension is the need for assistance in viewing it. However, let us consider a hypothetical scenario where you must go to a particular place. Upon arrival, you glance at your watch. The point is, you can never be present at a location without there being a time.

It is more difficult to grasp the concept of a universe without time than it is to comprehend one with time. Therefore, for the time being, time serves as the fourth dimension.

There are still other models that define the fourth dimension in distinct ways. Fortunately, scientists are making concerted efforts to validate what precisely the fourth dimension is.

## Perspective Twenty-One

# Would We Exist Without Time?

Time underpins all aspects of our lives. Were time a non-existent concept, it would follow that our lives would never come to fruition as we would be barren of the opportunity to live.

Yet, it is noteworthy to mention that time is intrinsically bound to all dimensions to enable their existence.

It is plausible to conjecture that existence is time, and vice versa; that time is existence. If we consider time as a dimension, it could very well be the most consequential dimension of all, as in its absence, no other dimension may be conceivable.

## Perspective Twenty-Two

# What is the Fastest Speed in the Universe?

In the realm of physics, matter is broadly defined as any entity that possesses mass and occupies space, thereby exhibiting volume. The physical properties of matter make it an object of great fascination for scientists and non-scientists alike. One of the most intriguing aspects of matter is its ability to travel at astonishingly high speeds, often in relation to the Earth. However, it is worth noting that matter can only approach, but never surpass, the remarkable speed of light. This universal constant travels at an astonishing 299,792,458 meters per second, and represents a fundamental limit on the velocity of any object in the universe. As postulated by Albert Einstein in his special theory of relativity, light's unparalleled speed makes it an incredibly powerful force in the cosmos, one that continues to captivate and inspire us to this day.

## Perspective Twenty-Three

# What Was the Discovery of Special Relativity?

In the year 1905, a physicist by the name of Albert Einstein introduced a pioneering theory that revolutionized our perception of the world. Special Relativity, a fundamental pillar of modern physics, was a complex concept that required diligent study to be understood properly. However, at its core, it introduced two principles that were fundamental to our understanding of the universe: the constancy of the speed of light and the relativity of simultaneity.

The first principle postulates that the speed of light in a vacuum remained constant for all observers, irrespective of their motion. This meant that no matter how fast an observer was moving relative to a light source, they always measured the speed of light to be approximately 299,792 kilometers per second. The second principle challenged our conventional understanding of simultaneous events. In special relativity, events that were simultaneous for one observer might not be simultaneous for another observer in motion. Time and space are intertwined into a four-dimensional continuum, which was known as spacetime.

The theory also introduced the concept of time dilation (discussed later in this book), where time appeared to pass more slowly for an observer in motion than it does for a stationary observer. This effect was accompanied by a contraction in length, which meant that objects in motion appeared shortened along the direction of motion when measured by a stationary observer.

Finally, Einstein's famous equation, $E=mc^2$, demonstrated the equivalence of mass and energy. It revealed that energy (E) was directly proportional to mass (m) and was equal to the mass times the speed of light (c) squared. This equation highlighted the profound connection between mass, energy, and the speed of light, and was considered one of the most significant scientific breakthroughs of the 20th century.

## Perspective Twenty-Four

# What is the Stubborn Illusion or Dilation of Time?

Consider a scenario where you are informed that your aging rate is influenced by your motion, and moving faster could potentially alter the pace of time. This concept might seem bewildering, as well as sound impossible, much like how Albert Einstein's special theory of relativity confounded several scientists upon its publication. However, Einstein expounded that motion triggers time dilation, where an observer's clock ticks slower as their speed increases. While this may seem implausible at first, it is indeed true.

To better understand this phenomenon, let us indulge ourselves in a thought experiment where you are seated inside a high-altitude airplane which can travel at an unprecedented speed. As the plane gathers speed, reaching 25%, 50%, 75%, and eventually 99.999% of the speed of light, time starts to play tricks on you. Despite the plane's engines not working at full capacity yet, it begins to accelerate slower, since attaining the speed of light is impossible according to special relativity.

If you were to get up from your seat, walk across the floor to the front of the plane, and run, you would be breaking the speed limit. However, that is not possible in our universe, and you would have no choice but to slow down, since light is the fastest speed possible. Strangely though, you would not even realize that you have slowed down, as your cognitive processes would have also slowed down, making you feel like you usually do while running. However, because of this, you would be aging slower than those observing you from the Earth's surface.

To someone on Earth, you would appear to be moving in slow motion, and your clock would tick at a different rate than the clock on the surface of the Earth. Additionally, if you were to look out of the plane window, you would perceive the universe to be unfurling before you at a much faster pace than those on Earth. Those on Earth would also appear to be aging and moving faster than you.

Upon returning to Earth, you would observe that your friends have aged more than you, since all the while you were on the plane, you were aging at a different rate than them. The amount of time you spend traveling and your speed determines your time dilation. Thus, if you were to travel on the plane for 100 years at the correct speed, perhaps you would have aged only one month, while your friends on Earth would have aged 100 years. This is the essence of time dilation.

## Perspective Twenty-Five

## What Happens to Time at the Speed of Light?

As per the Special Theory of Relativity proposed by Albert Einstein, light is deemed to be the swiftest speed attainable in the universe. However, if one were to travel at the speed of light, the ramifications on the perception of time remain a matter of utmost curiosity and intrigue. Would time dilation occur, or would it cease to exist? The theoretical possibility of traveling at the speed of light translates into a reality where one's clock would cease to tick, and the aging process would become a thing of the past. In effect, light could take an aeon of 100,000 years to traverse the universe before being absorbed, yet it would remain in sync with the moment it was created. Hence, time is rendered nonexistent at the speed of light, leaving us to ponder over the fundamental nature of space and time.

## Perspective Twenty-Six

# Do We Ever Stop Moving?

Have you ever paused to ponder the extent to which we are in motion? Not merely on the Earth's surface, but through the vast expanse of space itself. As we peruse these lines, we are already in rapid rotation on the Earth. With each rotation, we are also orbiting the sun at an astonishing velocity. Further, as the Earth orbits the sun, our star is itself revolving around the center of the Milky Way galaxy. And as the sun traverses the Milky Way, the galaxy, along with countless others, is hurtling through space.

Yet, as far as we can tell, there may be even more profound movements that remain beyond our comprehension. Nonetheless, even when we are at rest, whether conscious or otherwise, we continue on an extraordinary odyssey among the stars.

# PERSPECTIVE TWENTY-SEVEN

## How Does Time Dilation Work In Our Daily Lives?

The discovery of time dilation stands as a remarkable achievement in the field of physics, unveiling a world in which the concept of universal time appears to be a mere illusion. In fact, everything in the universe operates on its own internal clock, ticking at a unique rhythm. As one approaches the speed of light, their internal clock slows down, including their thought process, resulting in an imperceptible dilation of time. Interestingly, in our daily lives, we are constantly approaching the speed of light on a minor level, even in seemingly mundane activities such as walking across the room. Consequently, there exists a faint dilation in our internal clock, leading to negligible changes in our perception of time. However, when observing someone else, the subtle differences in time dilation become more apparent. For example, if one could measure these changes, they would notice that their friend appears to be moving slightly faster than themselves due to the time dilation effect. On a much larger scale, we can consider the Earth's rotation, which completes a full turn every 23 hours and 56 minutes, and orbits around the sun every 365 days, 6 hours and 9 minutes. Even these seemingly small movements encompass a minor time dilation, highlighting the extent to which time is intertwined with the fabric of our universe.

## Perspective Twenty-Eight

## What is Gravity?

The Earth's gravitational force that keeps us firmly grounded on the planet is a fundamental concept in the field of physics. Gravity can be envisioned as a magnetic force that pulls objects towards the center of the Earth. Defined as the force that attracts physical objects towards each other, gravity is a weak force in nature. Every object, including humans, has a degree of gravitational force. However, due to the small magnitude of gravity, we are unable to sense or feel it.

Gravity is significantly stronger for massive objects like the Earth. The spherical shape of planets is a result of gravity, which keeps objects in a stable configuration and prevents them from going out of shape. In contrast, minor asteroids lack the strength of gravity, thereby failing to maintain a spherical shape, unlike the Earth or the sun.

The absence of gravity would have a significant impact on the Universe. Mountains on Earth would grow much higher, and the Earth itself would be distorted. The absence of gravitational force would also mean that the Earth would not orbit the sun because the sun would lack the gravitational force to attract it. Furthermore, the Milky Way galaxy might not exist as the stars would be unable to orbit the center. Consequently, the Universe would look and behave quite differently without gravity.

## Perspective Twenty-Nine

# What is General Relativity?

A decade after the advent of special relativity, Albert Einstein presented a new theory called general relativity, in 1915. This revolutionary theory of gravity considered the influence of both mass and energy on the curvature of spacetime. Unlike the traditional Newtonian gravity, which portrayed gravity as a force between masses, general relativity described that massive objects like planets and stars warped the fabric of spacetime around them.

Einstein's field equations were the mathematical basis of general relativity, describing the curvature of spacetime in relation to the distribution of mass and energy in the universe. These equations had a solution that described the gravitational field around various objects, and predicted phenomena such as time dilation as gravitational time delays.

One of the most groundbreaking predictions of general relativity was the bending of light in a gravitational field, known as gravitational lensing. This effect had been observed and confirmed through experiments, such as the bending of starlight around the Sun during a solar eclipse.

Another crucial prediction of general relativity was the phenomenon of time dilation in strong gravitational fields. Clocks in stronger gravitational fields ticked more slowly than clocks in weaker gravitational fields, as validated by experiments involving highly precise atomic clocks.

Furthermore, general relativity predicted the existence of black holes, regions of spacetime with gravitational forces so strong that nothing - not even light - could escape. Black holes had been indirectly observed through their effects on nearby matter and the detection of gravitational waves, which were ripples in spacetime caused by the acceleration of massive objects. This theory had successfully explained a wide range of gravitational phenomena and had withstood numerous experimental tests and observations.

## Perspective Thirty

# What is Spacetime?

The concept of spacetime held a fundamental position in the realm of modern physics, particularly in the context of Einstein's theories of relativity. Spacetime unified space and time as a four-dimensional continuum instead of treating them as separate and absolute entities. This integration was crucial for comprehending the connection between space and time in a way that accounted for the constant speed of light and the relative nature of motion.

In the context of special relativity, spacetime emerged as a fabric that combined the three dimensions of space, namely length, width, and height, with the fourth dimension of time. This unified framework described objects, events, and even the fabric of the universe. The geometry of spacetime was influenced by the distribution of mass and energy, as outlined in general relativity.

The curvature of spacetime near massive objects, such as stars or planets, led to the gravitational effects we observed.

Large objects caused spacetime to curve around them, influencing the motion of other objects, including the paths of light rays. This curvature was what we perceived as gravity. In regions with strong gravity, such as near a black hole, spacetime was dramatically warped, leading to time dilation and gravitational lensing.

Spacetime had been crucial for our understanding of the cosmos and had been confirmed through many experiments and observations, displaying its role in the fabric of modern theoretical physics.

*Albert Einstein, the pioneer of Special and General Relativity.*
Photo source: Pixabay.com

## Perspective Thirty-One

# What is the Gravitational Time Dilation?

Albert Einstein's theories of relativity propose that the dimensions of space and time are not independent entities but are unified in a four-dimensional framework known as spacetime. In this framework, massive objects such as planets and stars warp the fabric of spacetime around them, influencing the motion of objects and leading to the phenomenon of gravitational time dilation.

Gravitational time dilation is a consequence of the impact of gravity on the fabric of spacetime. According to general relativity, time slows down in regions of stronger gravitational fields. This effect is most pronounced near massive objects, where the gravitational pull is stronger. When an observer moves closer to a massive object such as a star, they experience a slower passage of time compared to an observer farther away.

One of the ways to observe gravitational time dilation is by using clocks. Clocks placed within a stronger gravitational field, such as on the surface of a massive planet, tick more slowly than clocks in a weaker gravitational field, such as those in space. This phenomenon has been experimentally confirmed with highly precise atomic clocks. For example, atomic clocks on Earth's surface run slightly slower than the identical clocks on satellites in orbit due to the difference in gravitational strength.

The intertwining of space and time in the fabric of spacetime demonstrates how gravity influences not only the motion of objects but also the very perception of time itself. Gravitational time dilation continues to play a crucial role in our understanding of the relativistic effects in strong gravitational fields.

## Perspective Thirty-Two

# Is Time Travel Possible?

After an extensive exploration of the vast expanse of the known universe, we are now confronted with the quintessential inquiry that has captivated human imagination for centuries - is time travel a believable phenomenon? The concept of time dilation, which is an intrinsic aspect of our quotidian life, is a testament to the fact that we are always, in a sense, time-traveling. Furthermore, even the slightest of gravitational forces can induce changes in time.

If we were to hypothetically construct a starship that could transport humans to far-off destinations within a single lifetime, we may be able to exploit the bending of spacetime and time itself to our benefit. However, it is worth mentioning that the mere act of building satellites or spacecraft is equivalent to inventing time machines, as the very existence of our planet and everything within it is predicated on time travel.

## PERSPECTIVE THIRTY-THREE

# WHAT IS A BLACK HOLE?

The mysterious black holes continue to fascinate the scientific community and the general public alike. These cosmic entities are among the most perplexing phenomena of the Universe that we know of, and have only recently been captured in an image for the first time.

A black hole is a region in the fabric of spacetime where the gravitational pull is so intense that not even light can escape. This concept emerged from Einstein's general theory of relativity, which proposed that gravity is a result of curved spacetime caused by the presence of mass and energy. When a massive star exhausts its nuclear fuel, it may undergo a gravitational collapse, creating a point of infinite density known as singularity, surrounded by an invisible boundary called the event horizon.

The event horizon marks the point of no return around a black hole. Once an object crosses this boundary, it is inexorably drawn into the black hole's gravitational field, and no signal or information from that object can be transmitted to an external observer. The event horizon's size is proportional to the black hole's mass; larger black holes have larger event horizons.

Black holes can be detected indirectly through their gravitational effects on nearby matter, such as the bending of light known as gravitational lensing, and the emission of X-rays from hot gas that falls into the black hole. Black holes can range in size from stellar-mass black holes formed by the collapse of massive stars to supermassive black holes found at the centers of galaxies, including the Milky Way's center, with masses equivalent to millions or even billions of suns.

*One of many various artist's impressions of how a black hole may look.*

Photo source: Pixabay.com

LUKAS WINKELMANN

## PERSPECTIVE THIRTY-FOUR

# WHAT IS THE QUANTUM WORLD?

In the previous perspectives of this book, we explored the vastness of the universe and the concept of time in great depth. However, we have yet to delve into the complexities of the universe on a smaller scale. This is where the quantum world comes into play.

The quantum world is a realm of physics that is governed by the principles of quantum mechanics, which is a fundamental theory that describes the behavior of matter and energy at extremely small scales. At this scale, we encounter atoms and subatomic particles, which are the building blocks of everything we see around us.

Although the quantum world challenges our classical understanding of reality, it has been incredibly successful in explaining microscopic phenomena and has practical applications in many technologies, including transistors, lasers, and quantum computing. There is so much to explore in the quantum world that it could easily take another book to cover it briefly. Nonetheless, physicists continue to investigate this fascinating realm of physics, seeking a deeper understanding of its true fundamental nature.

## PERSPECTIVE THIRTY-FIVE

# WHAT IS QUANTUM MECHANICS?

The fundamental theory of quantum mechanics provides a comprehensive framework for understanding the behavior of matter and energy at the smallest scales. Originating in the early 20th century, this theory has brought about a paradigm shift in our perception of the microscopic world, revolutionizing the way we comprehend the basic constituents of the universe.

At its core, quantum mechanics rests on a handful of key principles, each of which presents a unique perspective into the workings of the quantum world. Let's delve deeper into each of these principles:

**Wave-particle duality:** This principle unveils the intriguing fact that particles, such as electrons and photons, exhibit both particle-like and wave-like properties. This duality is encapsulated in a mathematical expression called the wave function, which describes the probability of finding a particle at a specific location.

**Quantum superposition:** This principle asserts that particles can exist in a multitude of states simultaneously, and only collapse to a single state when measured. This challenges the classical notion of definiteness, where particles were believed to possess a definite position and momentum.

**Entanglement:** When particles become entangled, their quantum states become correlated to the extent that the state of one particle is directly linked to the state of another, regardless of the distance between them. Changes to the state of one particle instantaneously affect the state of the other, even if they are separated by vast distances.

**Uncertainty:** This principle is embodied in Heisenberg's uncertainty principle, which asserts that certain pairs of properties, such as momentum and position, cannot be precisely determined at the same time. The more accurately one property is measured, the less accurately the other can be determined.

While quantum mechanics has already proven its mettle in explaining the behavior of particles on the microscopic scale, its implications for the nature of reality and its challenges to classical knowledge continue to fascinate scientists. It forms the cornerstone of modern physics and is a crucial component of our understanding of the universe at its most fundamental levels.

## Perspective Thirty-Six

# What is Inside a Black Hole?

In the previous perspectives, we acknowledged the caution required when contemplating speculative ideas. Yet, speculation can also be a powerful tool that instigates new and imaginative concepts, leading to the discovery of actual truths. The interior of a black hole is one of the most enigmatic and speculative subjects within astrophysics. The fundamental equations of general relativity, which describe the curvature of spacetime around a black hole, predict a point of singularity at its core. A singularity is a region in which gravitational forces become infinitely strong, and the density becomes infinitely high, defying our current understanding of physics.

The point of singularity is enclosed by an event horizon, which defines the boundary beyond which nothing - not even light - can escape the gravitational pull of the black hole. As a result, the interior of a black hole remains concealed from direct observation, giving rise to the term "black hole." These objects appear black and invisible against the cosmic background because no light or information can travel from inside the event horizon to the outside.

The existing physics framework fails to explain the singularity. This is where even general relativity reaches its limits. The extreme conditions in this region suggest the necessity of a unifying theory of quantum mechanics and gravity - a theory of quantum gravity - to describe the behavior of matter and energy within a black hole. However, no such theory exists, and the nature of the singularity remains a mystery.

Lukas Winkelmann

Perspective Thirty-Seven

# Is There a Difference Between Quantum Mechanics and General Relativity?

Quantum Mechanics and General Relativity are two great theories in physics, each addressing distinct aspects of the physical universe. They were both independently developed and have been immensely successful in their respective subjects. Yet they operate under different principles and mathematical frameworks, and there is currently no complete theory that merges the two.

Quantum Mechanics primarily deals with the behavior of matter and energy at the microscopic scale. It introduces probabilistic phenomena, wave-particle duality, and the concept of quantum superposition. It has been highly successful in explaining the behavior of particles such as electrons and photons and is fundamental to our understanding of atomic and subatomic processes.

On the other hand, General Relativity is a theory of gravity and spacetime. It describes gravity as a curvature of spacetime caused by the presence of both mass and energy.

General Relativity has also successfully explained phenomena on cosmic scales, such as the motion of planets, the bending of light in gravitational fields, and the existence of black holes.

Challenge arises when attempting to link these two theories in extreme conditions, such as those found at the center of a black hole. In these situations, both quantum effects and strong gravitational fields come into play, requiring a theoretical framework that unifies Quantum Mechanics and General Relativity. There are many ongoing efforts to develop such a theory, often referred to as a theory of quantum gravity, and researchers explore various approaches that are known as string theory and loop quantum gravity. Until a unified theory is established, the realms described by Quantum Mechanics and General Relativity remain distinct, with each theory providing insights into different parts of the natural world.

## Perspective Thirty-Eight

# What is String Theory?

As you know, two leading theories are fundamental to the Universe: General Relativity and Quantum Mechanics. These two theories have experienced a fair amount of testing through the years. Although some may debate, these theories sound accurate at the moment.

However, both of these theories are mysteriously different. They more or less need something to unify them together.

String theory is a theoretical framework in physics which aims to provide a unified description of all fundamental forces and particles in the universe. Unlike the traditional particle physics, where point-like particles are the basic building blocks, string theory states that the fundamental entities are tiny, vibrating strings. These strings can vibrate at a range of frequencies, and the various vibrational modes correspond to different particles.

The theory emerged in the late 20th century as physicists attempted to unite quantum mechanics with general relativity and create a consistent framework that could describe the behavior of particles in both the quantum and gravitational realms. One of the features of string theory is its potential to naturally incorporate gravity into the quantum description, which will address a long-standing challenge in theoretical physics.

String theory also predicts that there are extra dimensions beyond the familiar three dimensions of space and one dimension of time. These extra dimensions are most commonly compactified, or easier said, "curled up," at incredibly small scales, making them not foreseeable in our everyday experiences. The existence of these extra dimensions allows string theory to obtain a variety of particles and forces in a more unified manner.

Today, there are several different versions of string theory. Each version describes different string structures and interactions.

However, despite its promising features, string theory has faced challenges and criticisms, including the lack of experimental verification and the existence of multiple solutions. Ongoing research aims to refine and develop string theory further, and it remains an active and influential area of theoretical physics with a great potential of revolutionizing our understanding of the fundamental nature of the universe. However, as long as there is an evidential debate in science, we still need more understanding.

# Reflection

This is where this branch of science may end for now. String theory is only one sprout of several departments on the Scientific Tree of Discovery (discussed in the final perspective of this book).

This tree is constantly growing and will not stop growing within our human time. Everyone may have an opportunity to add to the Scientific Tree of Discovery.

One day, when you read these perspectives, a single branch of discovery may have doubled or tripled in size.

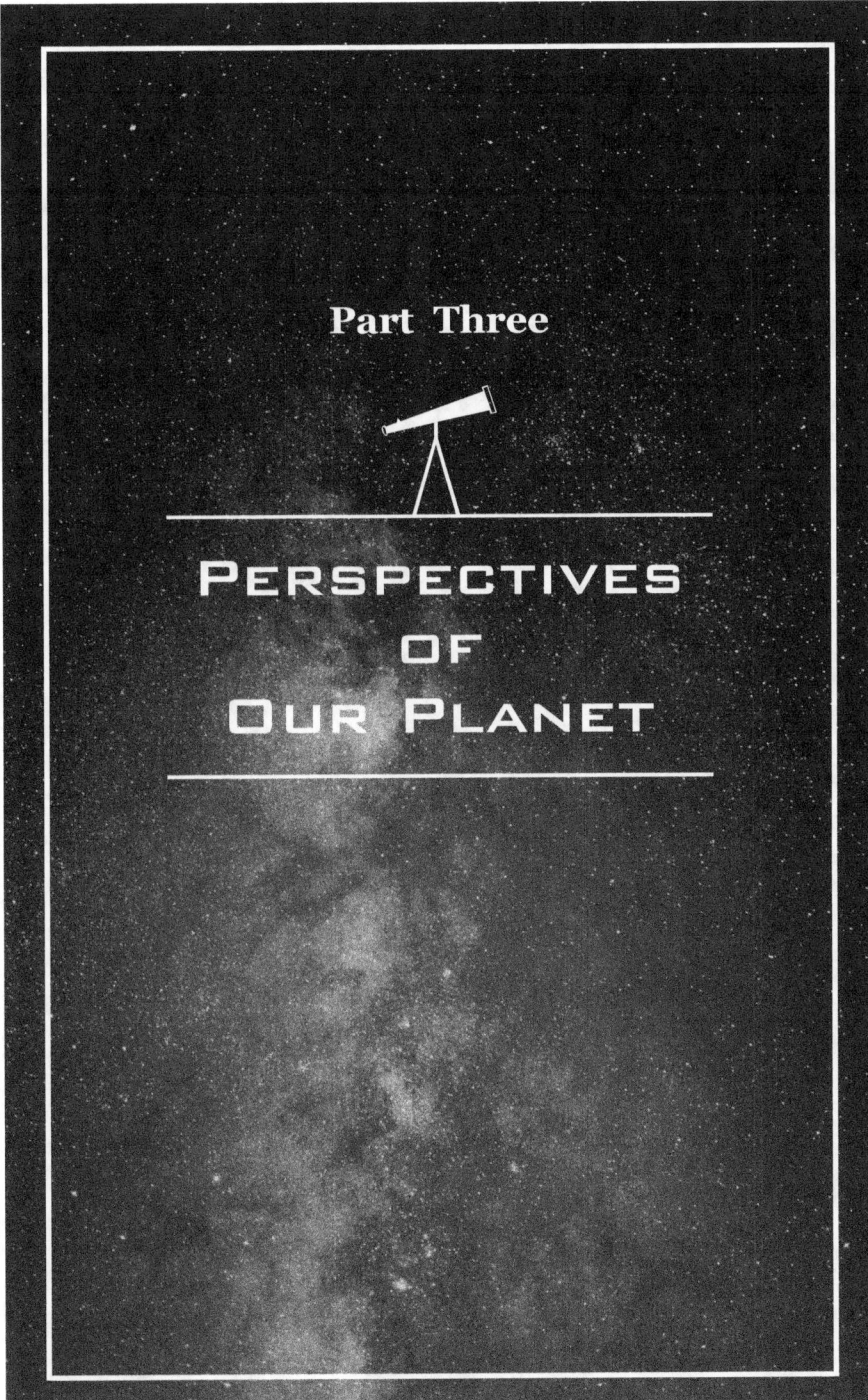

## Perspective Thirty-Nine

# Why Do So Many People Respect the Stars?

*"Only in the darkness can you see the stars."*
-Martin Luther King, Jr.

Have you ever wondered why so many people hold the stars in the sky in such high esteem? Is it because of our innate fear of the dark, and the stars act as cosmic traffic cones guiding us through the night? Or, perhaps it's the result of generations of ancestral beliefs that saw the stars as symbolic representations of our very purpose in life. But what if there's a deeper, more profound reason for our admiration of the stars? Could it be that we instinctively feel a connection to the cosmos, a sense of belonging to something far and beyond? As we continue to explore and collaborate across various platforms, we move one step closer to unraveling the mystery of why so many of us hold the stars in such high regard. The answer lies out there, waiting to be discovered.

## Perspective Forty

### Do Our Eyes See Everything?

*"Equipped with our five senses – along with telescopes and microscopes and mass spectrometers and seismographs and magnetometers and particle accelerators and detectors sensitive to the entire electromagnetic spectrum – we explore the universe and call the adventure science."*
-Edwin Powell Hubble

Have you ever wondered if what you see is everything that happens around you? If our eyes are capable of capturing every event that takes place in our lives? Well, the truth is, our eyes can only perceive a tiny portion of what happens around us.

The universe is full of electromagnetic radiation, which is displayed in what we call the electromagnetic spectrum. This spectrum includes Gamma Waves, X-rays, Ultraviolet Rays, Visible Light, Infrared Rays, Radar, FM, TV, Shortwave, and AM. Unfortunately, our eyes can only capture Visible Light, leaving us in the dark about the rest of the radiation.

But here's the intriguing part - some other organisms can distinguish beyond this region, giving them a broader perspective of the universe. This means that there is so much more happening around us that we are unable to perceive. So, if you're curious to know what else is out there, don't only search where your eyes can see...

## Perspective Forty-One

## Why Do Events Happen?

Each day, countless events unfold around us, shaping the world we inhabit. Often, we may be familiar with these events, yet understanding why they occur remains a mystery. As people, we desire to unravel the mysteries that surround us, but this quest for knowledge can be hindered by the distractions of everyday life. Unfortunately, our brains are not wired to multitask, and attempting to do so can lead to chaos and confusion. Despite this, many of us continue to believe that we can juggle multiple tasks with ease, leading to decreased productivity and increased stress. Imagine if we could devote our attention to just one event for a week - what insights could we glean about the world around us? By focusing on the present and embracing the mysteries that surround us, we can cultivate a newfound appreciation for the wonders of life.

## Perspective Forty-Two

# How Advanced is Our Civilization?

The quest to determine our position on the scale of civilization advancement is a fascinating yet complex one. We are limited by the fact that we can only truly understand our own society, leaving us without any other intelligent civilization to compare ourselves to. The possibility of more advanced societies existing beyond our reach makes it even more intriguing, yet frustratingly elusive. The truth is, we may never know how advanced we truly are relative to others, as our ability to comprehend a society more advanced than our own is not guaranteed. In essence, the question of our level of development remains a tantalizing mystery that continues to elude us.

## Perspective Forty-Three

# What is the Survival Rate of the Human Species?

Have you ever wondered whether we truly care about our environment? Or whether we have managed to overcome the global separations that have plagued our species for so long? And what about our survival rate as a species? Do we have what it takes to endure the challenges that lie ahead?

But here's something even more intriguing to ponder: what if an advanced alien society were to exist and look back at our human civilization? What would they think of us? And, what would we think of their existence?

If we truly want to be regarded as one of the most successful species on this planet, perhaps we should realize that our actions today will determine what future generations will think of us. So, let's strive to make a positive impact on the world and ensure a brighter future for all.

## How Much is the Earth Cared For?

*"It suddenly struck me that that tiny pea, pretty and blue, was the Earth. I put up my thumb and shut one eye, and my thumb blotted out the planet Earth. I didn't feel like a giant. I felt very, very small."*

-Neil Armstrong

The question that looms over us all is, do we truly value our planet as our sole habitat? And in doing so, do we strive to maintain a cohesive and unified society? With a diverse populace occupying the Earth, there are those who revere the natural environment and those who disregard it. But the truth remains, if we are to truly cherish the Earth and its ecosystem, we must also cherish every living being and their collective purpose.

## Perspective Forty-Five

# What is Earth Day?

Earth Day is a momentous occasion that happens only once a year, where people worldwide unite to showcase their unwavering support for the protection of our planet. This day is not merely a celebration of nature, but a call to action to safeguard Earth's climate. Participants take part in a range of activities, from planting new seeds and plants to clearing waste and debris from their surroundings. It's a celebration of our efforts towards creating a more environmentally sustainable society.

However, as we commemorate this day, we should also recognize a glaring issue - Earth Day only takes place once every 365 or 366 days. Shouldn't we be working towards making Earth a better place every day? While it may sound daunting to some, it's a necessary step towards ensuring a clean and safe environment for everyone.

Rather than viewing it as a chore, we should start looking at it as a way of life. It's not just about cleaning up after ourselves; it's about not making a mess in the first place. By doing so, we can help create a world that is clean, safe, and respected, not just for us but for future generations as well.

One common misconception is that we must "save" the Earth. However, the truth is that the Earth doesn't need saving. It will still be here. Instead, we should focus on preserving life on Earth, including our own. Our actions have a significant impact on the environment and energy around us. By treating them negatively, we're ultimately treating ourselves unhealthily. On the other hand, by treating them positively, we're treating ourselves and the environment around us in a healthy way. It's a win-win situation for everyone involved.

*Earth rising over the moon's surface taken by Apollo 8 astronauts on December 24, 1968.*

Photo source: Unsplash.com

## How Much Impact Does the Earth Have On Our Lives?

*"What is the use of a house if you haven't got a tolerable planet to put it on?"*
-Henry David Thoreau

The profound impact of our planet, Earth, on the lives of countless individuals cannot be underestimated. From its natural wonders to its awe-inspiring mysteries, the Earth has captivated the hearts and minds of people for centuries. But what is it that draws us to this tiny world? Is it an innate attraction that we cannot resist?

As we explore this question, we must consider the natural inclination of humanity to cling to our home planet. For many, the Earth is a sanctuary that provides us with everything we need to survive. Yet, what if we were born on a starship, where the Earth was a distant memory? Would we still feel drawn to it as we do now?

Living on a spacecraft, one might be constantly on the move, never settling in one place for too long. In such a scenario, would the idea of home become obsolete? Perhaps it is our attachment to the Earth that makes some feel spoiled, taking for granted the many wonders of this planet that we call home.

As we ponder these questions, it becomes clear that the Earth has a profound impact on our lives in ways that we may never fully understand. It is up to us to appreciate and cherish what this planet provides us, for it is truly a magnificent place.

*The view of Earth, as seen by the crew of Apollo 17 in 1972, the last Apollo mission to the moon.*

Photo source: Unsplash.com

## Perspective Forty-Seven

## What is Living, and What is Not?

The mere existence of life is a phenomenon that leaves us awestruck. As humans, we are driven by the innate desire to understand the purpose behind our existence. But before we delve into that, let's first try to define what life really is.

If we look at the universe as a whole, we could consider stars to be living entities. They are born, they shine, and eventually, they pass, much like the life forms we know. In fact, one may even argue that the stars are the cells of the universe, responsible for its growth.

However, there is one critical difference that sets us apart from the stars - our capacity for emotions. It is our feelings that give us the power to bring thought and consciousness to our world. As the representatives of the world's brain, our existence is vital for the planet's survival.

This is why we, and all the other remarkable species of life, exist - to bring thought and consciousness to the universe, and to make thoughtful decisions that will shape the course of our world.

## Perspective Forty-Eight

# Our World in the Present

Transport yourself back in time to the mid-1900s, a time when space exploration was at its peak in the 20th century. Discover a world very different from our own, where the United States and the Soviet Union were locked in a fierce space race. With groundbreaking technological advancements from past world conflicts, we were finally able to explore the vast expanse of space.

The Soviets made history by engineering the first rockets that blasted into space, propelling the first human to orbit and back. Meanwhile, the United States sent brave astronauts to the moon and back, pushing the limits of human exploration.

Despite the incredible advances in space travel, we have only scratched the surface of the great cosmic expanse. The possibilities for exploration and discovery are endless. So why have we stopped at the moon?

We should be planetary explorers, driven by our innate

desire to discover the unknown. However, political incentives and funding shortages have hindered our progress. We must find a way to protest against these obstacles in a methodical manner, carefully selecting our battles and thinking ahead.

Imagine what we could achieve if we all learned to protest methodically and work towards a common goal. The possibilities are endless, and our world would be transformed for the better.

Methodical Thinking = More Advancement and More Improvement

*Some speculate that there was once life on Mars. Although some forms of microbes may still live within its deep sandy crust, this photo perceives a delicate planet compared to our own, underlining the importance of staying aware and preserving our native home, the Earth.*

Photo source: Unsplash.com

## Perspective Forty-Nine

## Lessons Taught and Lessons to be Prepared

*"It has been said that astronomy is a humbling and character-building experience. There is perhaps no better demonstration of the folly of human conceits than this distant image of our tiny world."*

-Carl Sagan

Imagine being in the vast expanse of space, suspended halfway between the Earth and the moon, spellbound by the extraordinary sight of our spherical home. From here, we can see the Earth in all its glory - a celestial jewel adorned by scattered clouds and a marbled tapestry of lush green and deep blue. It's a breathtaking spectacle, an embodiment of perfection that was meticulously crafted for "us". But, as we gaze at this heavenly world, we can't help but ponder on our society and ourselves. Let's delve deeper into this train of thought.

We are the people of planet Earth... intelligent, brave, and adventurous. This planet is what we may always call home.

Think of all our politics, religions, sciences, and more. Our species appears to be beaming with life and many more significant possibilities.

Our personalities depend on our moods. Our moods depend on how we relate to one another. Relationships are critical to our species' survival.

The human industrialization of our world changed the way we existed more than a century ago. Many extraordinary lives have been saved since then. Lives that may have never had a chance to live until now.

No matter your race, your religion, or the country you were born in, we are all people, just with different manners. Each culture displays kindness differently. So, why do we still argue about this and can't just get along?

Perhaps the answer lies right in front of us. Humans of different societies have only sometimes been presented to other cultures. Therefore, we get offended when we see different mannerisms because we think they are a sign of disrespect. However, to other cultures, it is a sign of respect. Relationships with societies worldwide have been disparaged because of the misinterpretation of one another.

Of course, we are all biased to some of our own beliefs and traditions. We all have different ways of showing our thoughts and manners, but they result in the same demeanor to express kindness...

In today's era of technological advancements, global warming and climate change have become household terms. Yet, there are still those who remain skeptical of their existence. However, the reasons behind this disbelief are understandable. Those who advocate for climate change often indulge in activities that contribute to the problem, such as driving gasoline-powered cars, taking frequent flights, and promoting rocket technology for exploration.

Even the claim that electric cars are environmentally friendly is relative at best. This situation raises intriguing questions about the true nature of climate change and the ways in which we can truly combat it.

Did you know that most power plants still rely on non-environmentally friendly resources? This means that the power that fuels electric cars is not always as eco-friendly as you might think, thus defeating the purpose of using an electric car. But that's not all - electric cars can also be a hassle to use. They can take hours to charge, and planning your trips around charging stations can take away from the spontaneity of a fun road trip...

Luckily, there is always hope. Newer designs of electric cars foresee a future where they will not be such a hassle to use. In fact, there will most definitely be a moment where charging an electric car will be just as fast as putting gasoline into a gasoline powered car. And besides that, the day that all energy becomes clean will also be the day that electric cars become a clean use of transportation. Therefore, the future of electric cars is exciting, and these inventions can revolutionize transportation, and save millions of dollars worldwide because of their ever lasting energy source. So it is most important to remain in support of the great technological advancements made with the production of electric cars. They will one day be completely clean.

...

In the realm of environmentally-friendly power production, nuclear power has already made its mark. While it's true that nuclear power is a cleaner, safer alternative to other forms of energy, the memory of past disasters still looms large in the minds of many. As a result, nuclear power has often been dismissed as a viable option.

...

After assessing several factors, skepticism about the existence of climate change still remains. However, a proposed hypothesis must be tested to arrive at an accurate conclusion. Conducting experiments is a potential avenue to explore. The results of such experiments can provide valuable insights to the ongoing climate crisis. Nonetheless, it's crucial to note that many have already conducted experiments and reached the conclusion that climate change is real. If your experiments lead you to believe otherwise, it's essential to contact a reputable scientist for further research. Your findings may potentially contribute to scientific breakthroughs. However, it's important to bear in mind that without reproducible experiments providing evidence for your discovery, your claims may be viewed with skepticism.

As per the results of numerous authenticated experiments including my own, it is scientifically evident that climate change is an actual phenomenon. To mitigate its adverse effects, it is imperative to cease all fossil fuel power sources without delay. Now, although this does seem to be what has to be done, it always makes me a bit sad to say.

Through my personal experiences, I have had the opportunity to interact with individuals who work exclusively with fossil fuels, thereby contributing to the emission of non-environmentally friendly gasses. These people are kind and compassionate. However, the notion of completely eliminating fossil fuels can be perceived as a threat to their livelihoods, and as a result, the issue of climate change can become complicated and even political.

At the beginning of this book, science was acknowledged to be the sole domain that all nations of the world could collectively agree upon, devoid of any religion or politics. However, in the present times, we are witnessing a conflation of science and politics that is challenging this fundamental belief. Such an unfortunate development warrants a closer inspection.

The correlation between climate change and politics is a complex and multifaceted concept. For instance, advocating for the discontinuation of fossil fuels, although a viable solution to mitigate climate change, has social and economic implications that must be weighed and addressed. By me stating that we have to get rid of all fossil fuels is like stating that I want to ruin the lives of the people who I have always treated with respect and cared for. And without those jobs for those people, they might end up becoming homeless, which leaves me dreadfully sad at their unfair misfortune. This societal imbalance can lead to stagnation for some and further impoverishment for others, creating a daunting challenge for any government. Furthermore, climate change poses an additional challenge, which can further complicate the situation.

Let us delve into an alternative perspective of climate change's impact on society. In contemporary times, we are constantly bombarded with news headlines featuring climate change or global warming, which invariably convey a a pessimistic outlook about our planet. As we listen to scientists warning us about the impending doom and the ticking clock to our extinction, it raises a pertinent question: How does this affect our society? Let us explore this intriguing topic in greater detail.

While these news headlines aim to convey a significant message, their pervasive bleakness often leaves me yearning to silence the television. This raises a pertinent issue, as the incessant deluge of such despondent reports may culminate in a society steeped in negativity and despair.

In relaying messages to the public, it would be prudent for the news to abstain from negative framing. My intention is not to assert myself as a pushy teenager with lofty demands (although I may sound it, and if so, I'm sorry...we can chat about that later after the book is over), but rather to encourage novel solutions to the world's problems, given that conventional approaches have fallen short. Why not shift our focus to the potential positive outcomes of collective action for environmental preservation? We may have been being taught more about what not to do than what we should do in the past decade. Society may be more willing to listen if we are taught more about what to do. Let's look forward to the future and enjoy our moment. Let's never let the past ruin who we are and what we want to be.

In the previous paragraphs, I expressed my frustration towards scientists who predict our impending doom. However, I believe that it is imperative to take a different approach. Rather than simply lamenting the lack of public support for climate change, perhaps we can encourage people to focus on finding solutions and developing new clean energy sources. This would enable society and exploration to continue functioning as normal, without any negative social, financial, political, or religious impacts resulting from this crisis.

*Even though Climate Change may be a very important issue to resolve, it should not prevent exploration. Instead, they should work parallel. This astronaut is performing a spacewalk aboard the International Space Station, which orbits the Earth while allowing the astronauts on board to perform scientific research about the Earth and the Universe. It is exploration and action that really protect us from any incoming issue.*

Photo source: Unsplash.com

## The Scientific Tree of Discovery

People wonder about the stars.
Wondering why they are there.
Some say they are all too far
While others prefer to share.

When we discuss a subject so wide
Our minds are free to speculate.
Life and its meaning is hard to decide
So stars are what to emulate.

But at the moment, we're behind the dome
Trapped on this tiny dot.
The Earth is our only home
And all that life has got.

The Earth has a small effect
In a vast cosmic set.
We must treat it with respect
And prevent the environment's threat.

I dream of the day we begin to voyage
And begin to work as a whole.
That will be the day we manage
To begin our important role.

In today's world, where we are constantly bombarded with information, it is intriguing to note that poems have a certain appeal that allows us to remember them easily, and even enjoy them more than traditional prose. If we were to imbibe poetry throughout our lives, would we be able to memorize things better? Poems, with their vivid imagery, offer a unique way of connecting with readers and enhancing their understanding of the world around them.

Now, let us take a step back and reflect on some of the concepts that we have explored in this book. We have delved into the inner workings of science, and the methodical steps required to gain a deeper understanding of the world. We have journeyed through time and space, exploring the vastness of the universe and the intricacies of the smallest particles.

It is fascinating to note that every scientist has a unique perspective, and it is their willingness to share their ideas that has led to the remarkable progress that we have seen in the field of science. Much like a family tree, science is a community of curious individuals, working together to unravel the mysteries of the universe.

For this final perspective, we will explore science in a different way - as something ordinary yet extraordinary, something that is alive and constantly evolving. Let us envision our scientific past as an evergreen tree, with branches reaching out in all directions, representing the many discoveries that have been made over the years. We shall call it "The Scientific Tree of Discovery."

Behold the colossal tree of discovery, with thousands of branches that have shaped the scientific past and present. Its roots signify the origin of our quest for knowledge, dating back to the ancient philosophers and scientists who first unearthed facts about our world, including its size and shape. This significant milestone in human history is a testament to the dedication of these early pioneers whose legacy continues to inspire us to this day.

The trunk represents the foundation of scientific understanding that has been systematically built upon by generations of scientists, ancestors, and stability. The branches that sprout from the trunk are the scientists who have built upon past discoveries to expand our knowledge even further, with smaller branches representing those who have continued to refine and improve upon existing ideas.

Each branch supports a lush canopy of evergreen leaves that represents our modern understanding of the universe, quantum mechanics, and the mysteries of life. Among these leaves, new bright green sprouts emerge, representing the latest theories and ideas that have yet to be proven. Among the thousands of them is String Theory. These sprouts need a branch to support them, and that branch could be one of us.

Occasionally, a sprout may not fully mature, but even these lost or fallen sprouts help to fertilize the formation of new shoots, increasing the likelihood of new discoveries. Science is forever evolving, and every new discovery brings us closer to understanding our world and improving our lives.

The Tree of Discovery is constantly growing, and anyone who has made a contribution, no matter how big or small, has a place on one of its branches. This tree is a symbol of our ongoing quest for knowledge and the remarkable achievements that have been made possible through the power of scientific exploration.

Regrettably, there are still individuals who display aversion towards the scientific tree of discovery. These people may feel disinclined to engage in exploration, inquiry, or experimentation. But, it is imperative to keep politics and religion separate from science. Science must remain impartial and devoid of political or religious influences. It is acceptable to hold beliefs, as they provide solace. I have them myself. However, we can maintain our personal beliefs to ourselves and personally hope that some scientific discoveries will change after more evidence and knowledge is gained in the future.

If the entirety of the readership of this book were to care for the environment, it would be insufficient since it is written in English. For this book to have a substantial impact, it should be translated into a dozen or more languages. One day, you may comprehend the perspectives of this book in a different, translated language. However, the readership of this book may only be about 0.01% of the human population. A large section of the population must have access to basic necessities such as housing, education, and accurate news before they can read and gain knowledge. Therefore, it is crucial to raise awareness about the state of our surroundings.

While books and resources on environmental protection are crucial, they likely will not be enough to mitigate the negative impact on the environment. However, we can make progress if we work together to discover innovative solutions that maintain the stability and enjoyment of our lives just as much as always, while also benefitting the environment, and the unity between nations, and individuality. By sharing our knowledge and skills, we can learn from each other and ensure a better future for all of us. Our collective effort towards this goal will be appreciated by humanity, and its future.

I dream of the day we begin to voyage
And begin to work as a whole.
That will be the day we manage
To begin our important role...

The day we truly begin to explore the stars will be the day we work as a whole, but not to the extent where we lose our great individuality. Everyone could have their names on the Tree of Discovery.

# Epilogue

My name is Lukas Winkelmann. I have written 50 Perspectives of the Cosmos, hoping to share my imagination of the cosmos and dedicate time to express our modern scientific language.

I had many reasons to write this book as a teenager. I felt that maybe, just maybe, I would have a tiny bit more freedom and lenience for errors in understanding and perspective while I'm young rather than in adulthood. Writing this book as a High School Freshman allowed me to view the world with a broader viewpoint before I grow up to a much different routine, with responsibilities outside of high school.

In recent times, I have personally noticed a separation between global societies. It is tough to agree upon something as a whole. Some believe that that is just the way we are. But I certainly don't think that.

Our world, and we as a species, are young, but already, we are advanced. We could be deep space explorers within our lifetime!

However, some others prefer to stay here on Earth. That is okay. But we should agree upon how to manage our planet well. Let's do more experiments, then allow others to do the same. If those disagree, be tolerant and listen because maybe we don't know enough about the topic yet...

If there is a debate about perhaps anything, we don't know the truth or obtain evidence to the systematic answer to the problem "yet." All we have to do is think more methodically during our lives.

With this book, I want to have a little part in opening our minds and possibilities of ourselves. Some have perspectives that disagree with science. It is okay to have your views on the world. Everyone should have the freedom to express their own views.

However, claims are different from perspectives. When someone states a claim, they need testable evidence to support it. Therefore, others can test and follow the same steps as the person who made the claim. This is not plagiarism. It is an improvement, as well as making sure it is true.

If their claim is invalid, it should be converted to a perspective. This can make some people rage with anger. People do not want to say that they are not right. People, including myself, try to aim to be correct. But that is not so easy to achieve, however...

I have personal thoughts about our society.

Perspectives are equally as significant as claims, if not even better. Perspectives are more open. People can have more imagination and more ideas. Perspectives could be the blossoms on the Scientific Tree of Discovery.

Imagine how great a world would be with only perspectives; no debates on who is right would be necessary, just acceptance... A world of views would be much more calming. I would no longer worry about my passion compared to my religion.

In a perfect world of perspectives, neither science nor religion rules the world. We, the humans, with our individual beliefs, are the ones who oversee it. Anyone could have the power to share their scientific thoughts with society. Anyone could have the ability to share their beliefs as a perspective to the organization.

I look forward to that world. That could be a safe world. One with a secure environment. One without the separations between borders. One without the labeling of each other. And one where we all follow our passions.

You and I could talk across seas without worrying about any prohibited issues. We could have the freedom to talk about anything and everything...

*"To me, it underscores our responsibility to deal more kindly with one another, and to preserve and cherish the pale blue dot, the only home we've ever known."*

-Carl Sagan,
Pale Blue Dot, 1994

## About The Author:
## Lukas Winkelmann

Lukas Winkelmann is a high school student whose passion for science and nature started in middle school amidst the COVID-19 pandemic. Despite the world being at a time of great separation, Lukas found hope in the influx of space missions and new voyages to the International Space Station. As space agencies and companies revealed new future technologies with the hope of transporting people to the moon and beyond, Lukas became a passionate supporter of astronomy and the exploration of the local galactic neighborhood. During this period, he spent much of his time watching documentaries on space travel and science, learning astrophotography, and developing his first understanding of physics and the universe. As schools began to reopen and the pandemic started to fade, Lukas founded clubs at his school to explore reaching beyond the atmosphere above and building other STEM technologies.

www.ingramcontent.com/pod-product-compliance
Lightning Source LLC
Chambersburg PA
CBHW050109230526
45470CB00004B/1754